This book belongs to

This book is made by

Color the  numbers

COLORING NUMBERS

COLOR THE NUMBERS

COLORING NUMBERS

COLOR THE NUMBERS

COLORING NUMBERS

COLOR THE NUMBERS

COLORING NUMBERS

COLOR THE NUMBERS

COLORING NUMBERS

COLOR THE NUMBERS

COLORING NUMBERS

COLOR THE NUMBERS

COLORING NUMBERS

COLOR THE NUMBERS

COLORING NUMBERS

COLOR THE NUMBERS

COLORING NUMBERS

COLOR THE NUMBERS

COLORING NUMBERS

COLOR THE NUMBERS

Count numbers 1 to 10

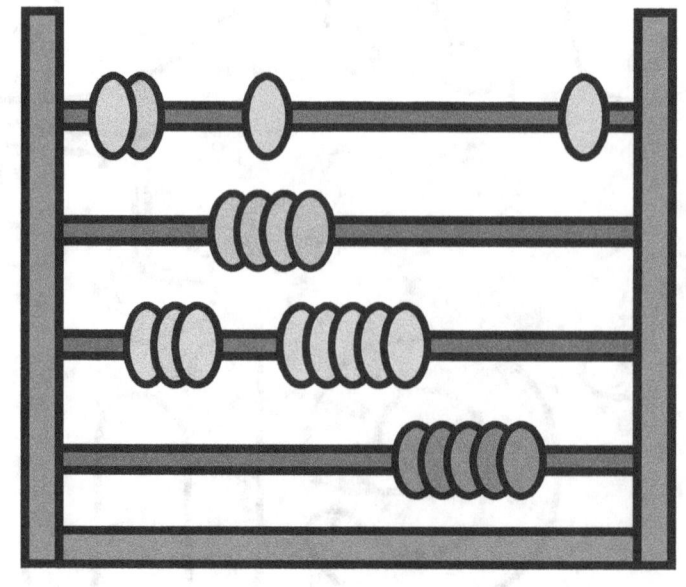

Name _____

Color the numbers indicated

7	
4	
3	
8	
2	
10	
5	

Name _____

COLOR THE NUMBERS INDICATED

4	
7	
8	
3	
10	
2	
6	

Name _____

COLOR THE NUMBERS INDICATED

8	
3	
7	
9	
1	
6	
2	

Name _____

COLOR THE NUMBERS INDICATED

1	
2	
3	
5	
4	
6	
8	

Name _____

COLOR THE NUMBERS INDICATED

5	
8	
6	
4	
1	
3	
2	

Name _____

Color the numbers indicated

10	
7	
8	
6	
9	
3	
2	

NAME _____   HOW MANY?

COUNT THE PICTURES AND WRITE THE NUMBER

NAME _____   HOW MANY?

COUNT THE PICTURES AND WRITE THE NUMBER

Name _____ HOW MANY?

COUNT THE PICTURES AND WRITE THE NUMBER

Name _____ COUNT AND MATCH

COUNT THE PICTURES AND MATCH WITH THE NUMBERS

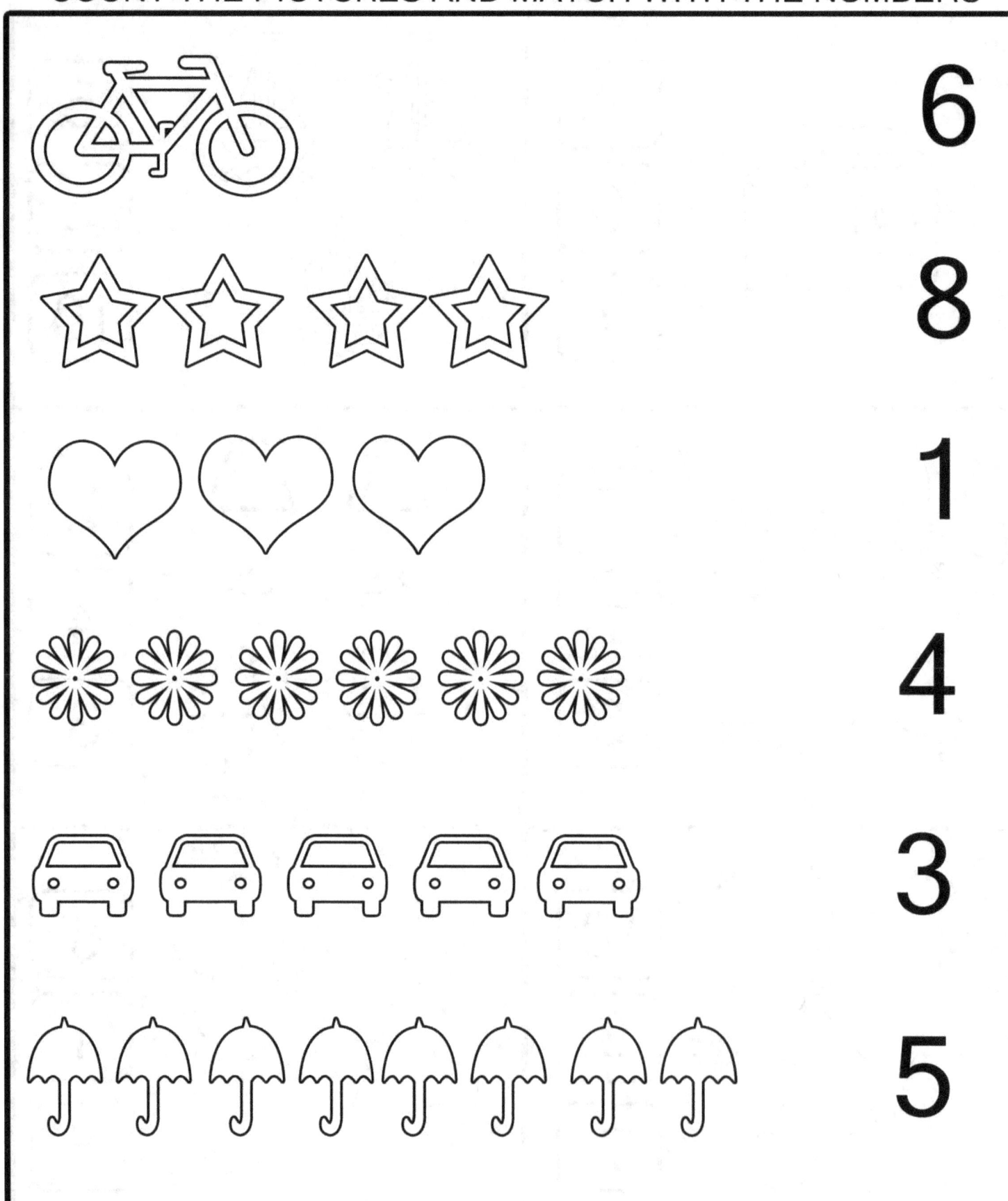

Name _____   COUNT AND MATCH

COUNT THE PICTURES AND MATCH WITH THE NUMBERS

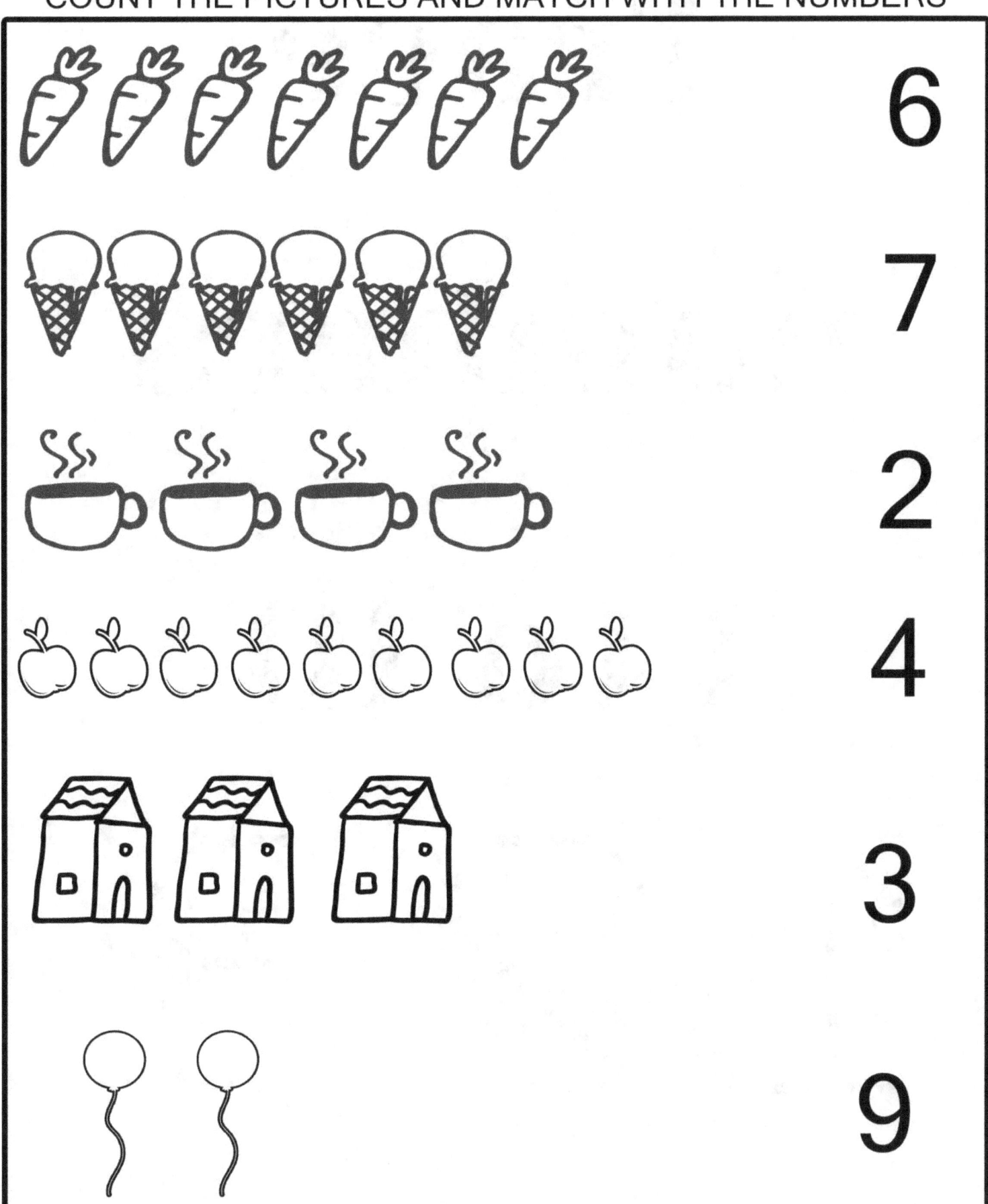

Trace numbers 1 to 10

NAME _____   WRITING NUMBERS

TRACE AND WRITE THE NUMBERS

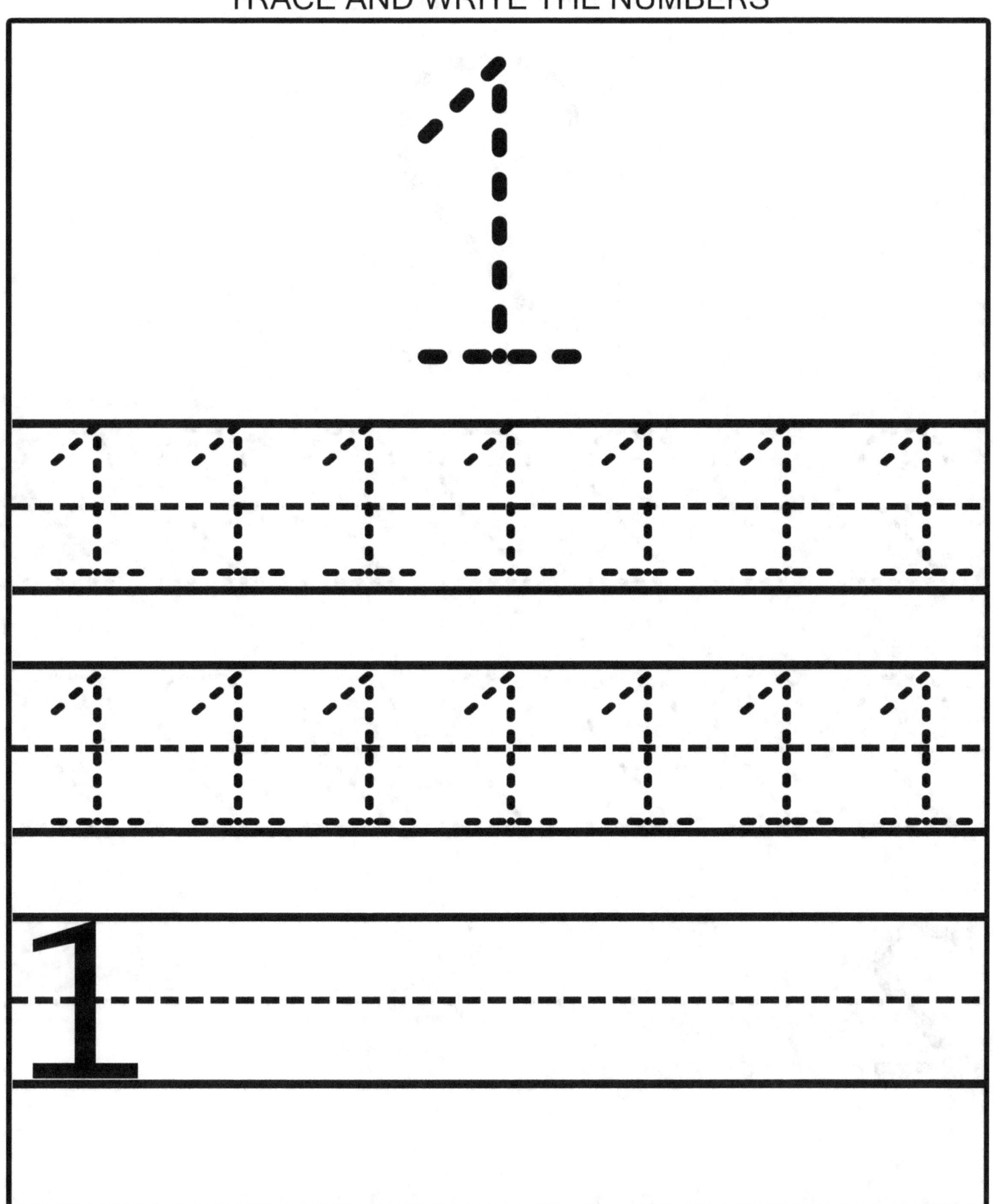

Name _____   WRITING NUMBERS

TRACE AND WRITE THE NUMBERS

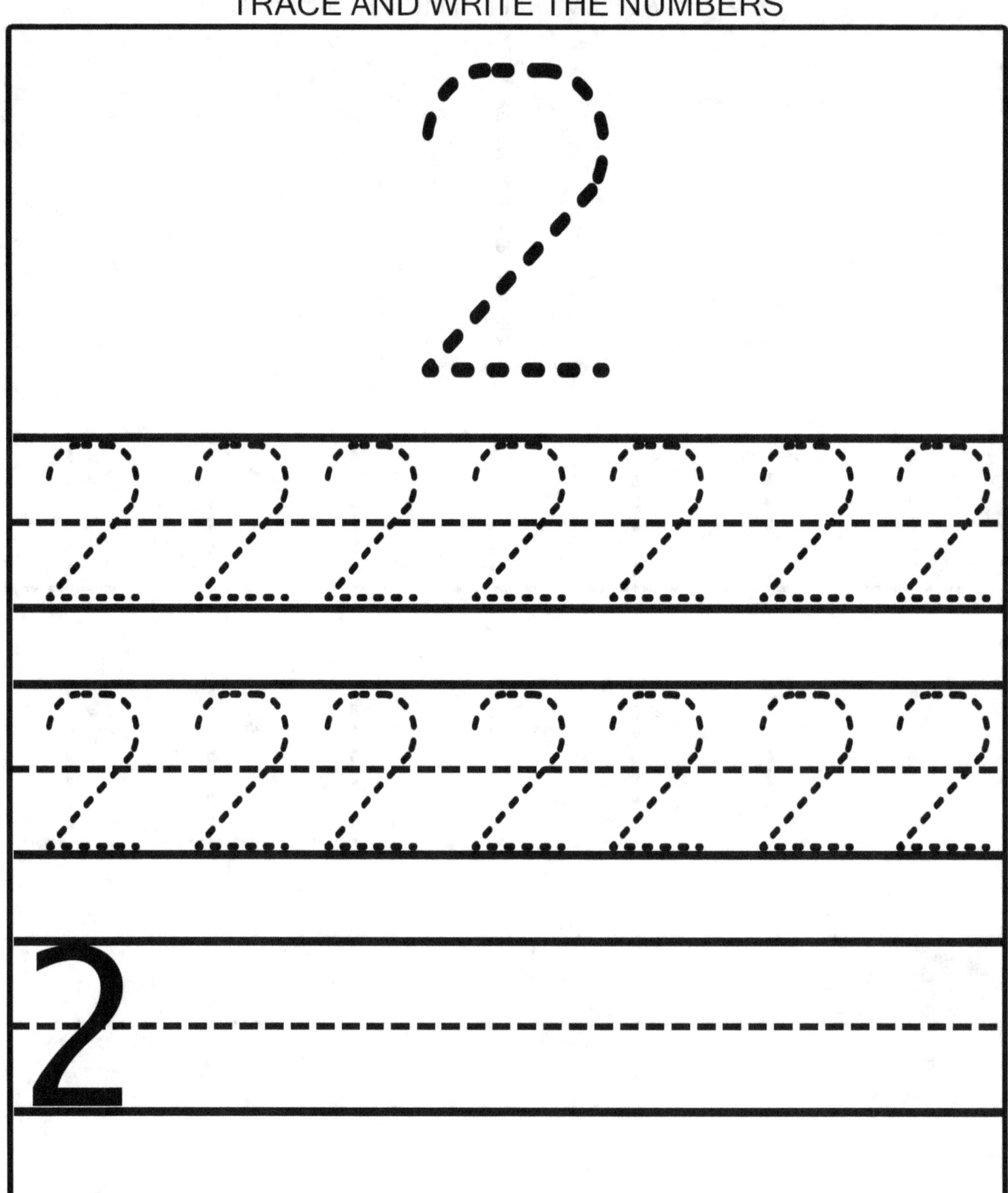

NAME _____   WRITING NUMBERS

TRACE AND WRITE THE NUMBERS

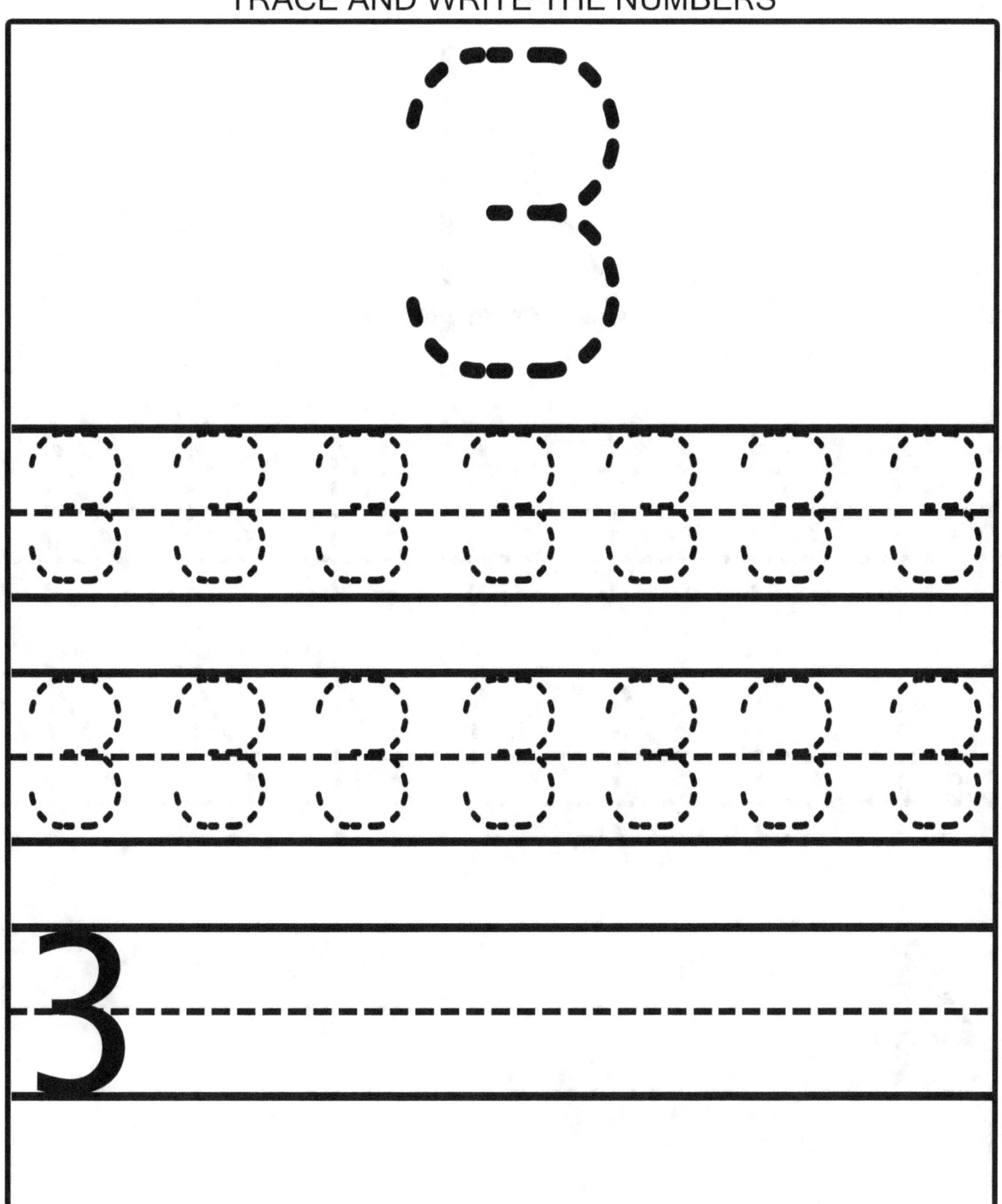

NAME _____ WRITING NUMBERS

TRACE AND WRITE THE NUMBERS

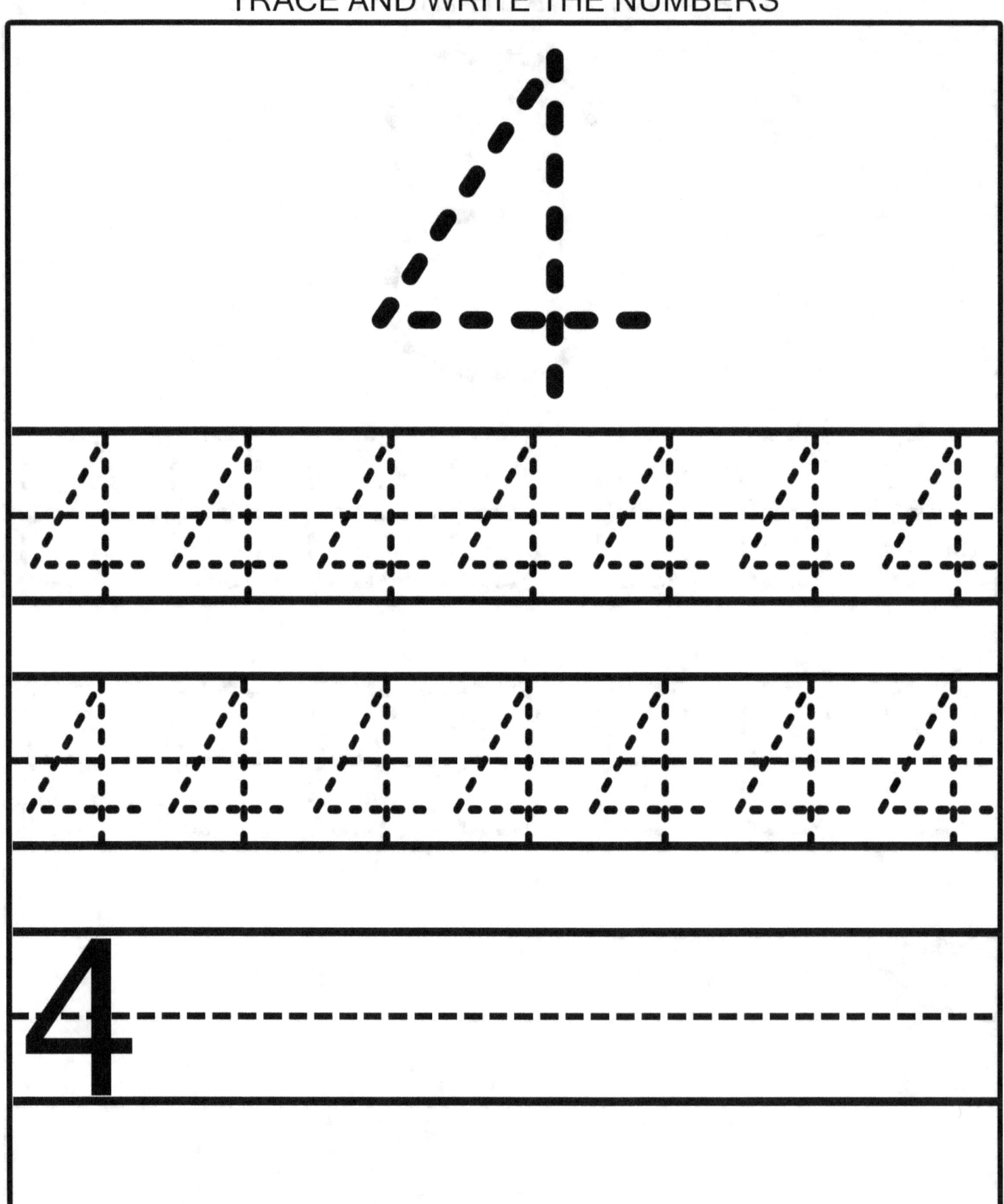

NAME _____ WRITING NUMBERS

TRACE AND WRITE THE NUMBERS

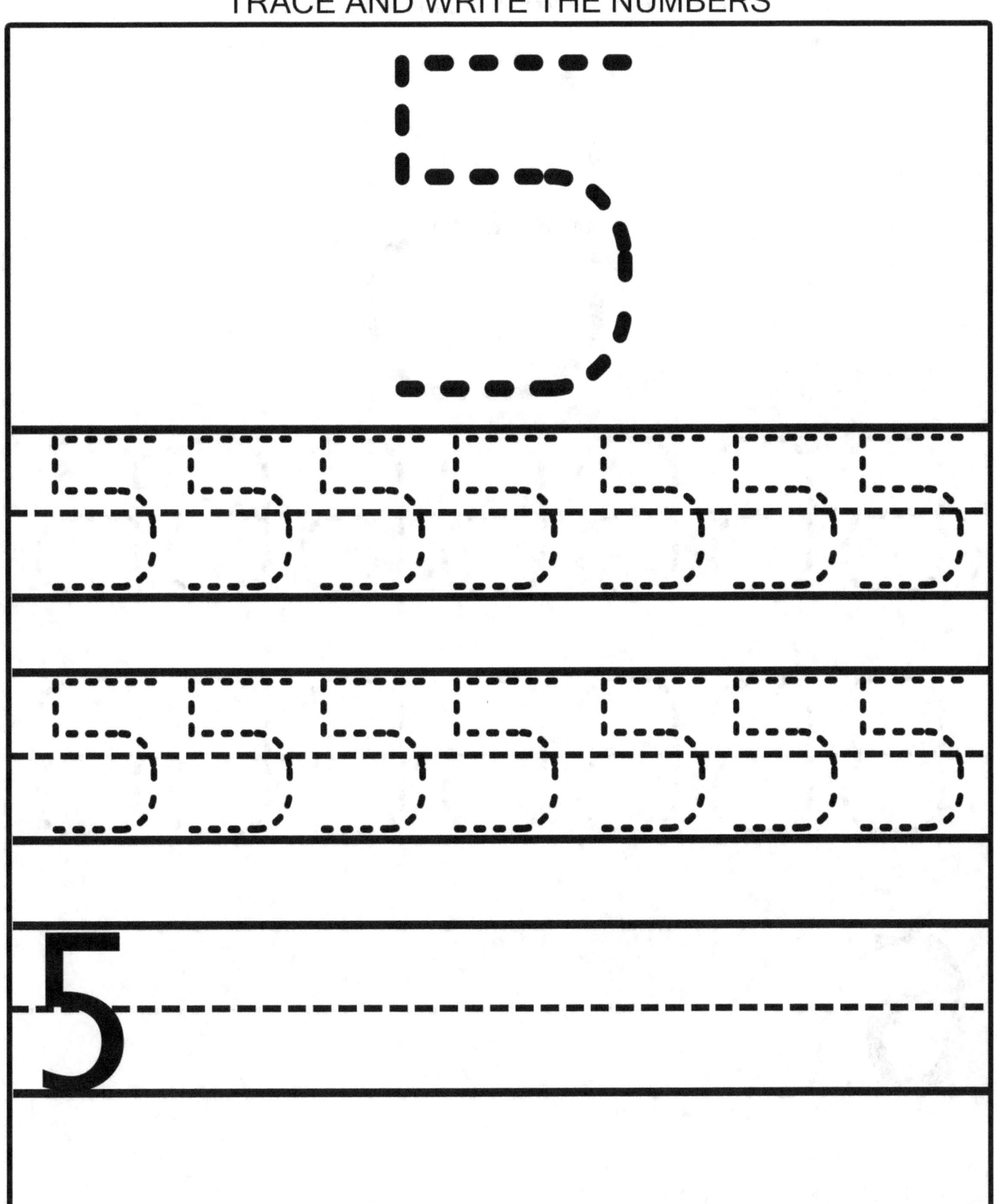

NAME _____ WRITING NUMBERS

TRACE AND WRITE THE NUMBERS

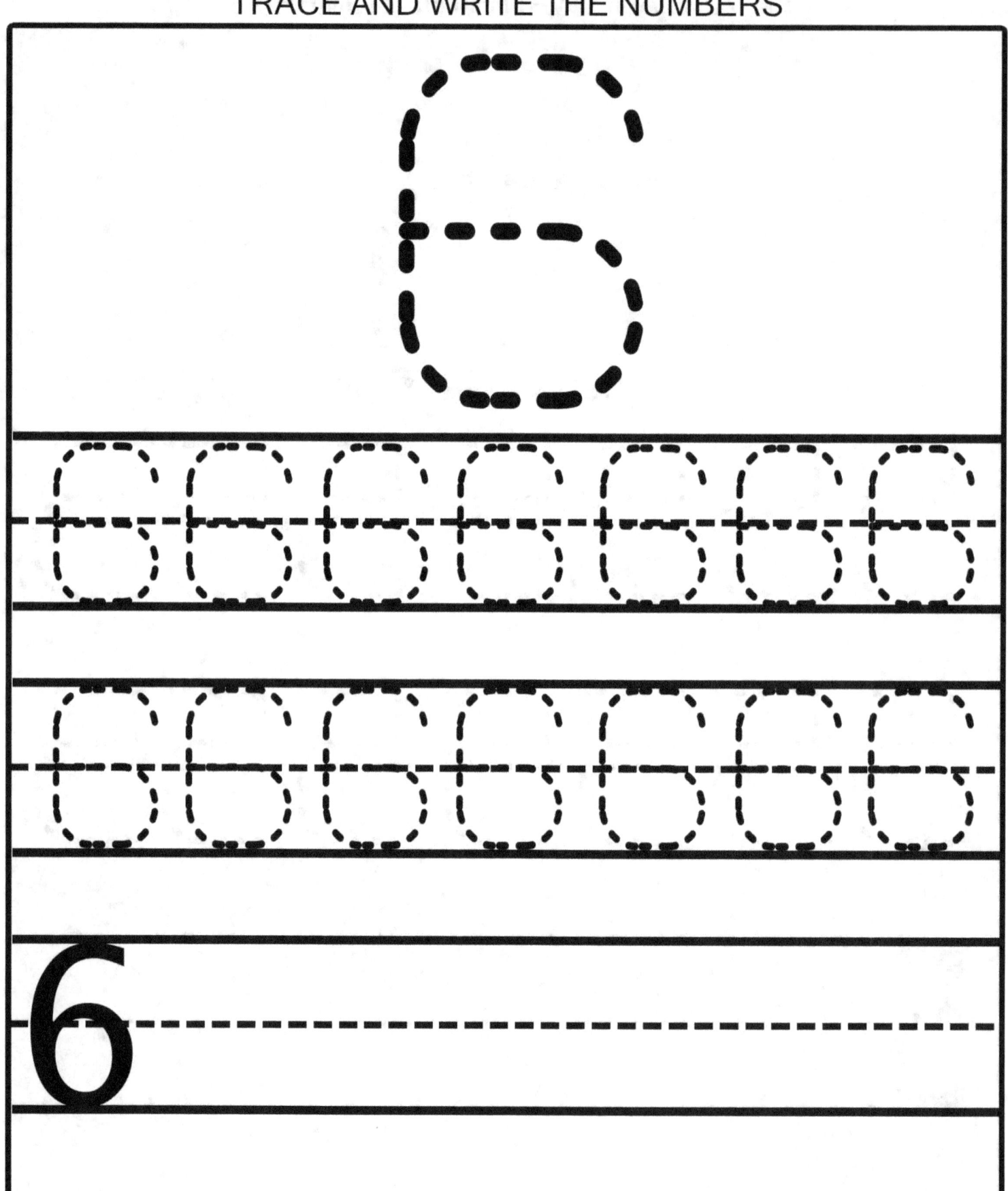

NAME _____ WRITING NUMBERS

TRACE AND WRITE THE NUMBERS

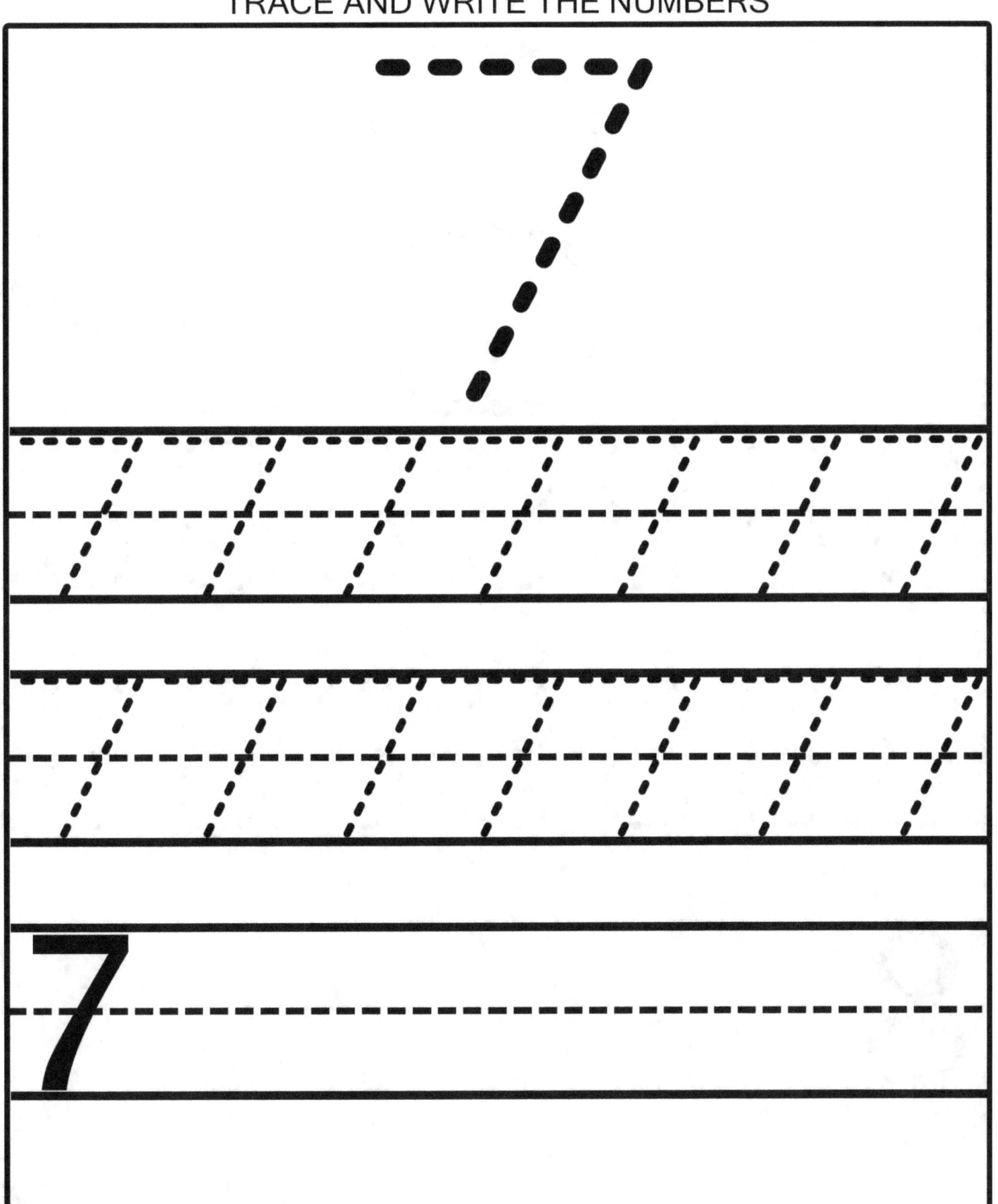

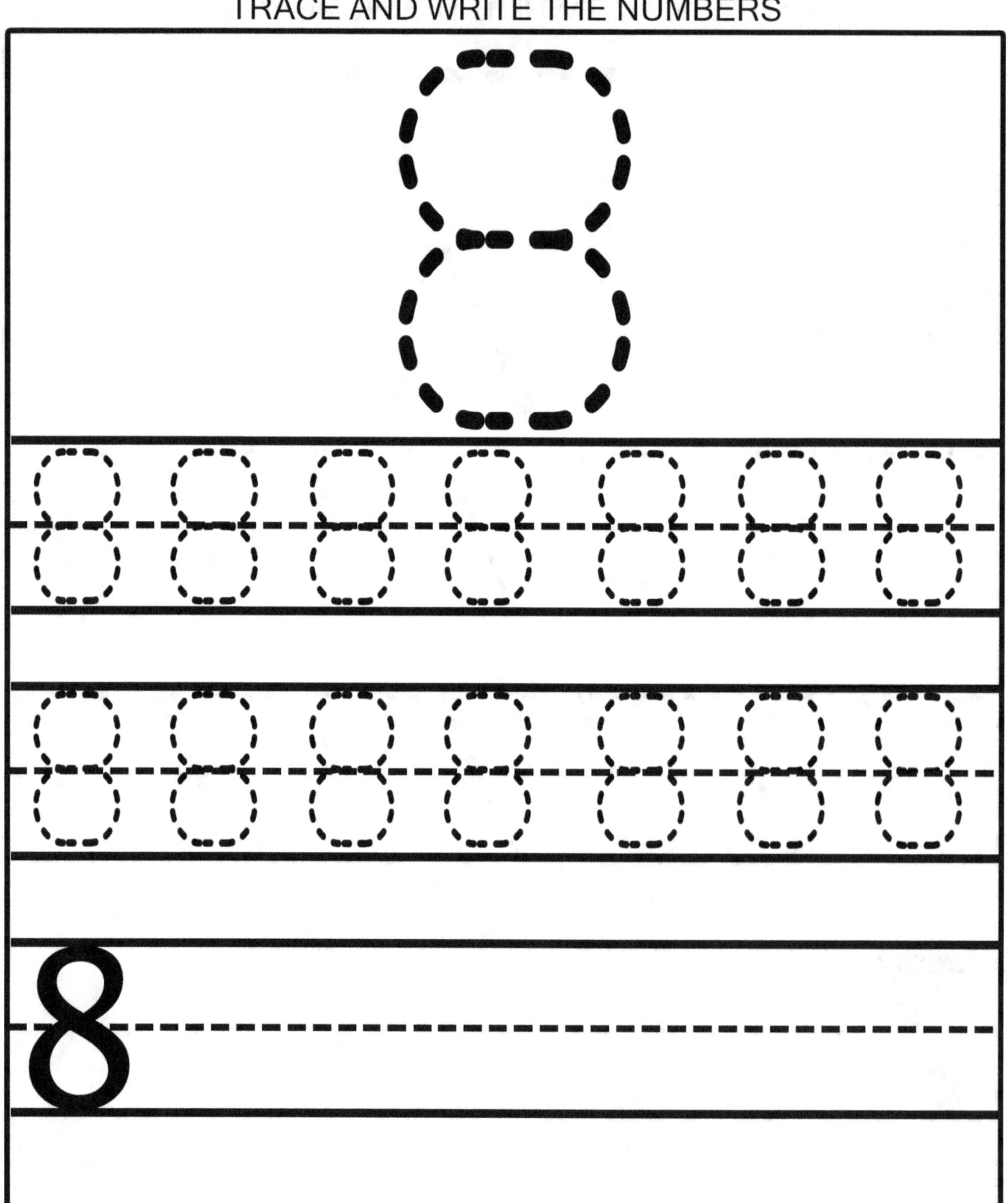

Name _____ WRITING NUMBERS

TRACE AND WRITE THE NUMBERS

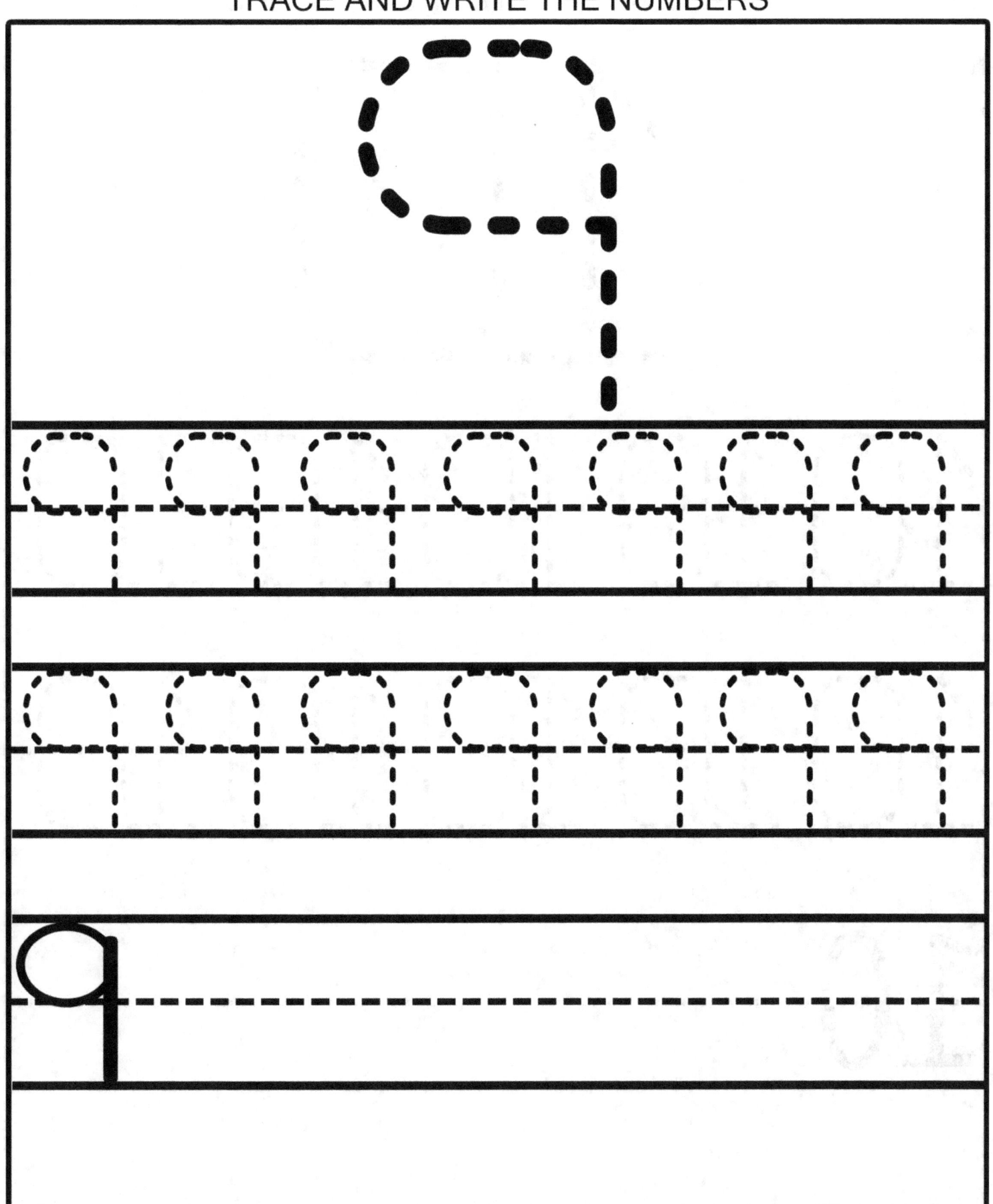

NAME _____ WRITING NUMBERS

TRACE AND WRITE THE NUMBERS

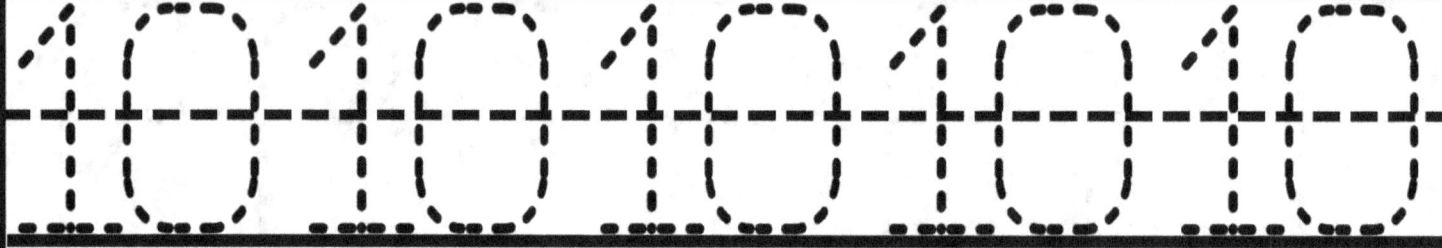

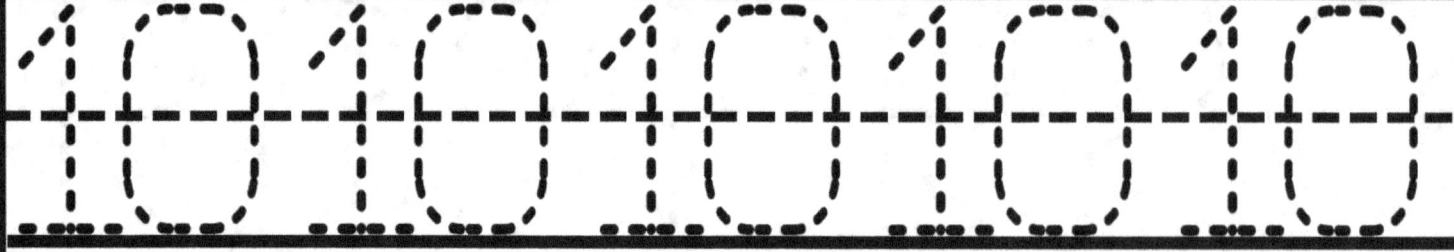

10

www.ingramcontent.com/pod-product-compliance
Lightning Source LLC
Chambersburg PA
CBHW080438220526
45465CB00009B/3333